Published by: AoPS Incorporated
                     15330 Avenue of Science
                     San Diego, CA 92128
                     info@BeastAcademy.com

ISBN: 978-1-934124-64-2

Beast Academy is a registered
trademark of AoPS Incorporated.

Written by Jason Batterson
Illustrated by Erich Owen
Additional Illustrations by Paul Cox
Colored by Greta Selman

Visit the Beast Academy website at www.BeastAcademy.com.
Visit the Art of Problem Solving website at www.artofproblemsolving.com.
Printed in the United States of America.
2022 Printing.

Become a Math Beast!
For additional books,
printables, and more, visit
BeastAcademy.com

This is Guide 5C in a four-book series:

Want more Beast Academy?
Try Beast Academy Online!

Learn more at BeastAcademy.com

Contents:

Lizzie
"The BOOkwOrm"
Can name every
dragOn species
On
Beast Island
(alphabetically)

Alex
"The Executive"
plans tO run fOr
city cOmptrOller
when he's Old enOugh
fOr public Office

Winnie
"The Firecracker"   Feisty!
GrOws 50 times
her Original size when angry!
(nOt really, but it's fun
to draw her that way)

GrOgg (me)
"The ˄ DenOminatOr"
    least commOn

ALter EgO:
FractiOn JacksOn!

Mr. Wriggles

Kraken
Shop Teacher

Favorite pattern?
Arrrrrgyle

Favorite holiday?
Arrrrrbor Day

Favorite element?
~~Arrrrrrgon~~
Gold

Fiona
Math Team Coach

Donated her hair to
"Braids for Mermaids"
this summer

Professor Grok
Math Lab
(full of booby traps)

"Calamitous Clod"

Constantly
captured by

Ms. Q.
Math Teacher

Spends a lot
of time
with Mr. A.

R&G
Campus Maintenance
Engineer(s?)

Let me ride
in their
golf cart
once!

Sgt. Rote →
Gym Teacher

Can bench press
three times
his own bodyweight!
(4 lbs.)

Welcome to Beast Academy!

This book is called the Guide.

There is also a separate Practice book with lots of problems you can use to sharpen your skills.

The Guide is written like a comic book.

In a comic book, whatever I say shows up in these bubbles. They're called comic balloons.

Here's one!

Each character has a different balloon color. This makes it easy to tell who is talking.

My balloons are purple!

The story is told in panels.

Panels usually have a rectangular frame around them...

...like this one.

8

Practice: Pages 6, 36, and 74.

Contents: Chapter 7

See page 6 in the Practice book for a recommended reading/practice sequence for Chapter 7.

Chapter 7:
Sequences

R&G
Next

Kanga doodle-doo

Guess what today is!

Ummm... Tuesday?

Nope!

Well, I guess it is Tuesday, but more importantly...

...it's new cereal day!

What's so special about new cereal day?

I get to solve all of the puzzles on the back of the new box!

Let's see what fun Monster Munchies has in store for us today.

SUPER SOLVER games
What Comes Next?!

Directions: Choose the square that best completes each pattern below.

Solve all five to earn a certified Super Solver badge!

SUPER SOLVER

Ooooh! "What Comes Next?"

I love these.

1.

What Comes Next?!

Puzzle number 1.

Shapes and colors. Hmmm...

What comes next?

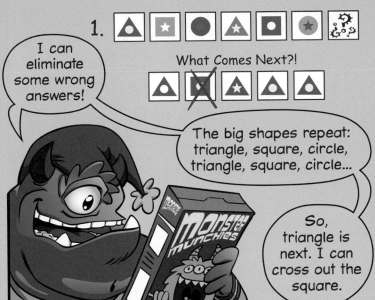

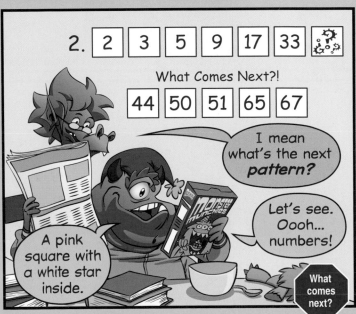

15

LOOK AT THE PATTERN UPSIDE-DOWN FOR THE ANSWER.

Ms. Q.
Sequences

What is a sequence?

Sequins are those shiny things Winnie has on her backpack!

I have them on one of my vests!

Sequence, Grogg! Not *sequins*.

Oh.

A sequence is a list of terms, usually numbers.

For example, 1, 3, 5, 7, 9 is a sequence of 1-digit odd numbers.

1, 3, 5, 7, 9

Some sequences go on forever.

For example, the sequence of positive multiples of 7 begins 7, 14, 21, 28...

...and goes on and on and on and on.

7, 14, 21, 28, ...

THREE DOTS AT THE END OF A SEQUENCE LIKE THE ONE ABOVE MEAN THAT THE PATTERN CONTINUES FOREVER.

Sequences that go on forever are called *infinite sequences*.

Sequences that end are called *finite sequences*.

Infinite: 1, 3, 5, 7, 9, ...

Finite: 1, 3, 5, 7, 9.

The order of the terms in a sequence matters.

For example, 2, 4, 6, 8 is a different sequence than 8, 2, 6, 4.

2, 4, 6, 8

8, 2, 6, 4

Most interesting sequences follow a rule, or make a pattern...

...like the sequence of primes...

Primes:
2, 3, 5, 7, 11, 13, ...

Powers of 2:
2, 4, 8, 16, 32, ...

...or the powers of 2, starting with 2.

Great. Let's find the next terms of a few sequences.

What are the next three terms in each of these sequences?

2, 4, 6, 8, 10, 12, __, __, __, ...

1, 4, 9, 16, 25, 36, __, __, __, ...

−1, 2, −3, 4, −5, 6, __, __, __, ...

Find the next three terms of each sequence.

21

Term: $2, 4, 6, 8, 10, 12, \underline{14}, \underline{16}, \underline{18}, \ldots \ 2n$
Position: $1^{st} \ 2^{nd} \ 3^{rd} \ 4^{th} \ 5^{th} \ 6^{th} \ 7^{th} \ 8^{th} \ 9^{th} \ \ldots \ n^{th}$

In the first sequence, each term is twice its position in the sequence.

So, the nth term is $2n$.

Term: $1, 4, 9, 16, 25, 36, \underline{49}, \underline{64}, \underline{81}, \ldots \ n^2$
Position: $1^{st} \ 2^{nd} \ 3^{rd} \ 4^{th} \ 5^{th} \ 6^{th} \ 7^{th} \ 8^{th} \ 9^{th} \ \ldots \ n^{th}$

In the second sequence, each term is the square of its position in the sequence.

So, the nth term is n^2.

Term: $-1, 2, -3, 4, -5, 6, \underline{-7}, \underline{8}, \underline{-9}, \ldots$
Position: $1^{st} \ 2^{nd} \ 3^{rd} \ 4^{th} \ 5^{th} \ 6^{th} \ 7^{th} \ 8^{th} \ 9^{th} \ \ldots \ n^{th}$

In this sequence, sometimes the nth term is n...

...other times, it's $-n$.

Term: $-1, 2, -3, 4, -5, 6, \underline{-7}, \underline{8}, \underline{-9}, \ldots$
Position: $1^{st} \ 2^{nd} \ 3^{rd} \ 4^{th} \ 5^{th} \ 6^{th} \ 7^{th} \ 8^{th} \ 9^{th} \ \ldots \ n^{th}$

We need to write an expression that is negative for odd values of n, but positive for even values of n.

How do we write an expression that works for both?

The terms in odd positions are negative...

...and the terms in even positions are positive.

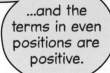

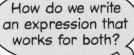

Can you find an expression that works?

$$(-1)^{99} = -1$$
$$(-1)^{100} = 1$$

Term: −1, 2, −3, 4, −5, 6, −7, 8, −9, ... $n \cdot (-1)^n$
Position: 1st 2nd 3rd 4th 5th 6th 7th 8th 9th ... nth

Subtractionacci

Subtractionacci is a pencil-and-paper game for two players.

Play

In a game of Subtractionacci, players take turns as the Picker and the Subtractor.

The Picker selects a two-digit number to be used as the first term of a Subtractionacci sequence.

The Subtractor selects a smaller number that is the second term of the sequence.

The sequence is completed by the Subtractor according to the following rules:

To find the next term in the sequence, take the two previous terms and subtract the smaller one from the larger one.
The sequence ends when a term in the sequence is larger than the one before it. Three examples are shown below:

Find a partner and play!

Picker chooses: 41	Picker chooses: 75	Picker chooses: 98
Subtractor chooses: 23	Subtractor chooses: 48	Subtractor chooses: 66
The third term is 41−23= 18	The third term is 75−48 = 27	The third term is 98−66 = 32
The fourth term is 23−18= 5	The fourth term is 48−27 = 21	The last term is 66−32 = 34
The last term is 18−5= 13	The fifth term is 27−21= 6	
	The last term is 21−6 = 15	

The last term is added to the Picker's score. In the three examples above, the Picker scores 13, 15, and 34 points.
The goal of the Subtractor is to choose a number that will give the Picker as few points as possible.

Players take turns as the Picker and the Subtractor. Pickers may not choose the same number twice.

Winning

The first player to score 100 points wins.

Strategies to Consider:

- As Picker, is it generally better to choose a large or a small number?

- As Subtractor, is it better to choose a number that is more than or less than half of the Picker's number?

- As Subtractor, is it better to choose a number that is more than or less than two thirds of the Picker's number?

- If the Picker chooses 89, what number can the Subtractor choose that gives the Picker just 1 point?

- The name of the game, Subtractionacci, is a reference to the Fibonacci sequence. Learn more about the Fibonacci sequence online or in the Practice book. What does this game have to do with the Fibonacci sequence?

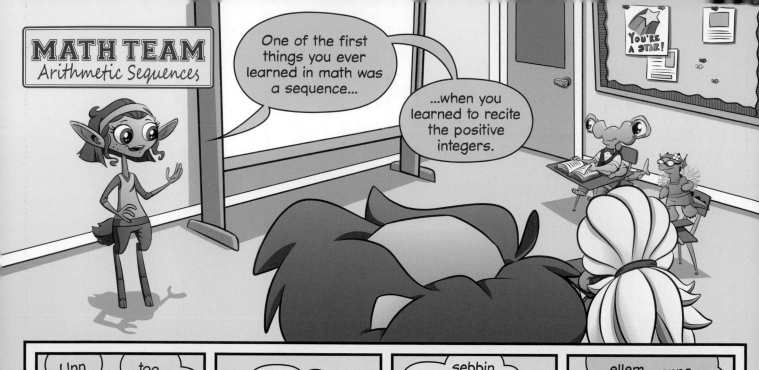

MATH TEAM
Arithmetic Sequences

One of the first things you ever learned in math was a sequence...

...when you learned to recite the positive integers.

Unn too fee...

...foe fye...

...sick sebbin ate...

...ellem menno P!

Later on, you learned how to skip-count.

Starting at a number, we always add the same amount to get the next term.

For example, the terms of this sequence start at 50 and count by 7's.

A sequence where we always add the same amount to get from one term to the next is called an **arithmetic sequence**.

How could you find the 100th term of this arithmetic sequence?

$$50, \ 57, \ 64, \ 71, \ 78, \ \ldots$$

What is the 100th term?

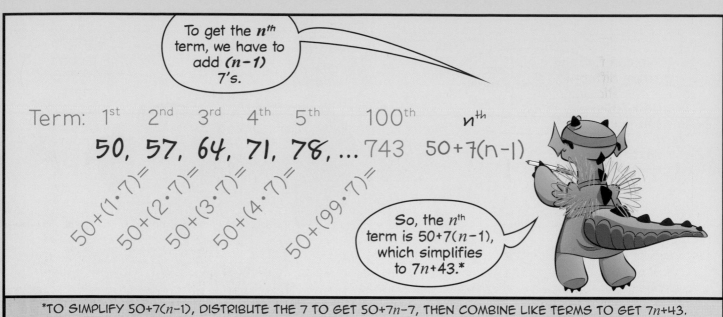

To get the nth term, we have to add $(n-1)$ 7's.

Term: 1st   2nd   3rd   4th   5th   100th n^{th}
50, 57, 64, 71, 78, ... 743 $50+7(n-1)$

$50+(1\cdot7)=$
$50+(2\cdot7)=$
$50+(3\cdot7)=$
$50+(4\cdot7)=$
$50+(99\cdot7)=$

So, the nth term is $50+7(n-1)$, which simplifies to $7n+43$.*

*TO SIMPLIFY $50+7(n-1)$, DISTRIBUTE THE 7 TO GET $50+7n-7$, THEN COMBINE LIKE TERMS TO GET $7n+43$.

Well done.

The amount you add is called the **common difference** of the sequence.

$$\overset{+7}{\frown}\ \overset{+7}{\frown}\ \overset{+7}{\frown}\ \overset{+7}{\frown}$$
50, 57, 64, 71, 78, ...

This sequence has a common difference of 7.

Find the common difference for each of these arithmetic sequences.

$199, 298, 397, 496, ...$

$13, 5, -3, -11, -19, ...$

$\dfrac{1}{6}, \dfrac{1}{4}, \dfrac{1}{3}, \dfrac{5}{12}, \dfrac{1}{2}, ...$

Try it.

30

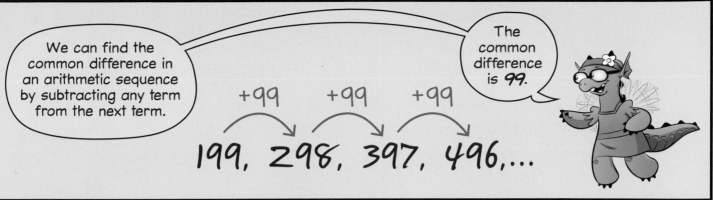

We can find the common difference in an arithmetic sequence by subtracting any term from the next term.

The common difference is **99**.

+99   +99   +99

199, 298, 397, 496,...

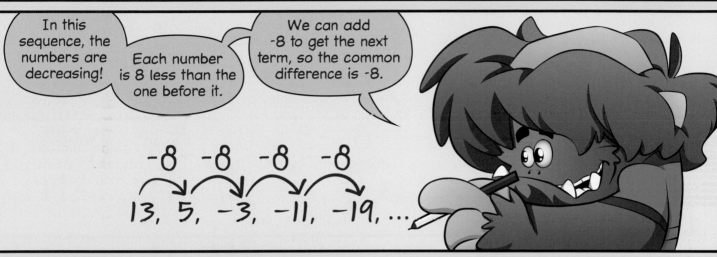

In this sequence, the numbers are decreasing!

Each number is 8 less than the one before it.

We can add -8 to get the next term, so the common difference is -8.

-8   -8   -8   -8

13, 5, −3, −11, −19, ...

This one looked hard at first...

...until I changed all of the fractions into twelfths.

The common difference is $\frac{1}{12}$.

$$\frac{1}{6}, \frac{1}{4}, \frac{1}{3}, \frac{5}{12}, \frac{1}{2}, ...$$

$$\frac{2}{12}, \frac{3}{12}, \frac{4}{12}, \frac{5}{12}, \frac{6}{12}, ...$$

$+\frac{1}{12}$ $+\frac{1}{12}$ $+\frac{1}{12}$ $+\frac{1}{12}$

Perfect. Use these common differences to compute the 21st term of each sequence.

+99   +99   +99

199, 298, 397, 496, ...

-8   -8   -8   -8

13, 5, −3, −11, −19, ...

$+\frac{1}{12}$ $+\frac{1}{12}$ $+\frac{1}{12}$ $+\frac{1}{12}$

$$\frac{1}{6}, \frac{1}{4}, \frac{1}{3}, \frac{5}{12}, \frac{1}{2}, ...$$

Try it.

31

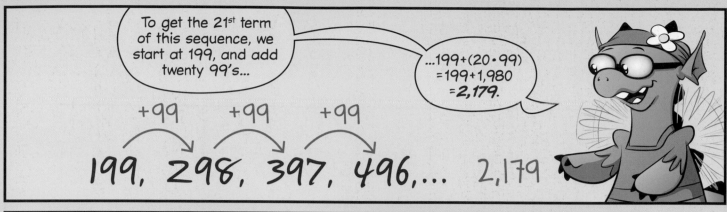

To get the 21st term of this sequence, we start at 199, and add twenty 99's...

...199+(20·99)
=199+1,980
=**2,179**.

+99 +99 +99

199, 298, 397, 496,...   2,179

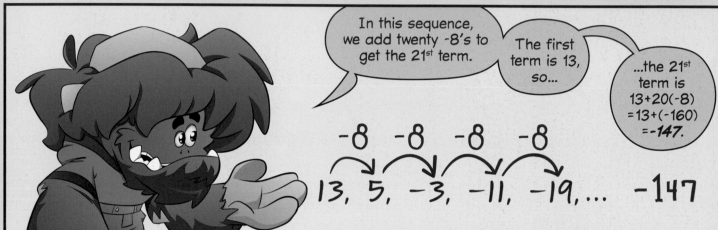

In this sequence, we add twenty -8's to get the 21st term.

The first term is 13, so...

...the 21st term is
13+20(-8)
=13+(-160)
=**-147**.

-8 -8 -8 -8

13, 5, -3, -11, -19,...   -147

And in this one, we add twenty twelfths to one sixth to get the 21st term.

$\frac{1}{6} + \left(20 \cdot \frac{1}{12}\right)$
$= \frac{1}{6} + \frac{20}{12}$
$= \frac{1}{6} + \frac{10}{6}$
$= \frac{11}{6}$ ...

...or $1\frac{5}{6}$.

$+\frac{1}{12}$ $+\frac{1}{12}$ $+\frac{1}{12}$ $+\frac{1}{12}$

$\frac{1}{6}$, $\frac{1}{4}$, $\frac{1}{3}$, $\frac{5}{12}$, $\frac{1}{2}$, ... $1\frac{5}{6}$

Great work.

In this next problem, you need to find some missing terms in the **middle** of an arithmetic sequence.

How would you begin?

10, ___, ___, ___, 54.

Try it.

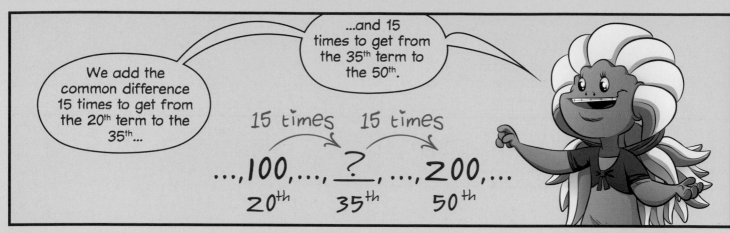

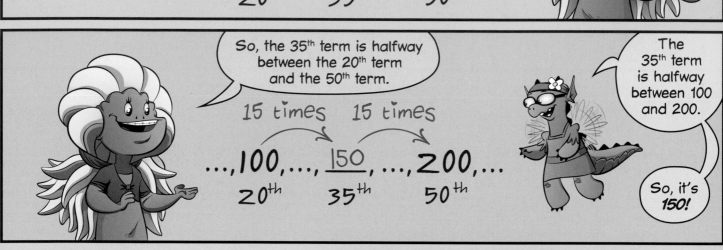

What are the next three?

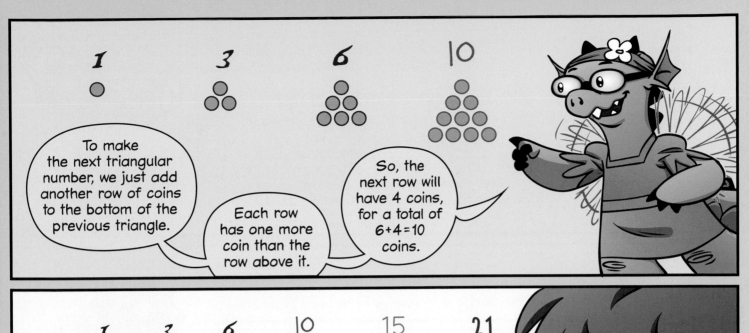

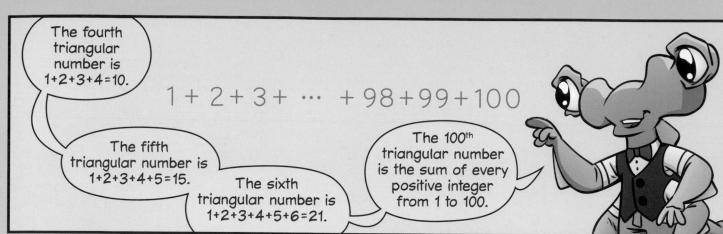

The fourth triangular number is 1+2+3+4=10.

$$1 + 2 + 3 + \cdots + 98 + 99 + 100$$

The fifth triangular number is 1+2+3+4+5=15.

The sixth triangular number is 1+2+3+4+5+6=21.

The 100th triangular number is the sum of every positive integer from 1 to 100.

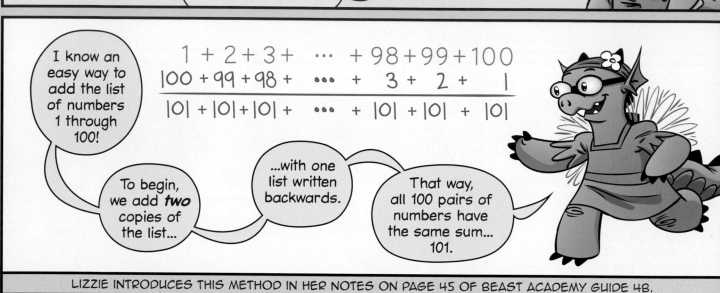

I know an easy way to add the list of numbers 1 through 100!

$$1 + 2 + 3 + \cdots + 98 + 99 + 100$$
$$100 + 99 + 98 + \cdots + 3 + 2 + 1$$
$$101 + 101 + 101 + \cdots + 101 + 101 + 101$$

To begin, we add *two* copies of the list...

...with one list written backwards.

That way, all 100 pairs of numbers have the same sum... 101.

LIZZIE INTRODUCES THIS METHOD IN HER NOTES ON PAGE 45 OF BEAST ACADEMY GUIDE 4B.

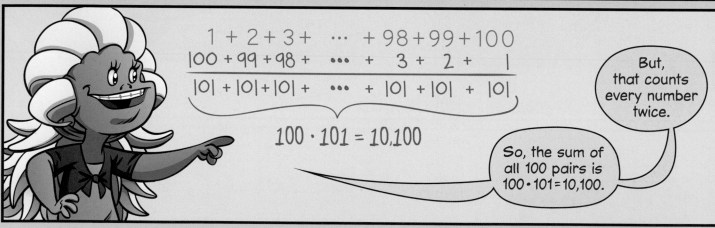

$$1 + 2 + 3 + \cdots + 98 + 99 + 100$$
$$100 + 99 + 98 + \cdots + 3 + 2 + 1$$
$$101 + 101 + 101 + \cdots + 101 + 101 + 101$$

$$100 \cdot 101 = 10{,}100$$

But, that counts every number twice.

So, the sum of all 100 pairs is 100·101=10,100.

$$1, \quad 3, \quad 6, \quad 10, \quad 15, \quad 21, \quad \ldots, \quad 5{,}050$$

$1^{st} \quad 2^{nd} \quad 3^{rd} \quad 4^{th} \quad 5^{th} \quad 6^{th} \qquad 100^{th}$

If the sum of *two* copies of the list is 10,100...

...then the sum of 1 copy of the list is 10,100÷2=5,050.

So, the 100th triangular number is 5,050.

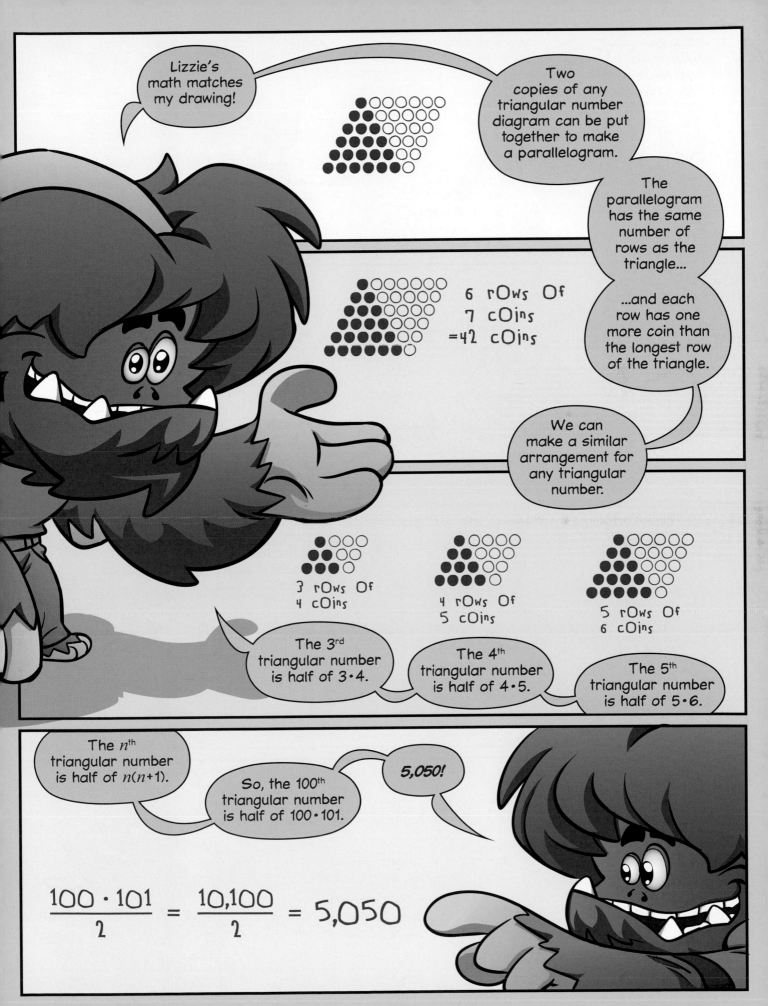

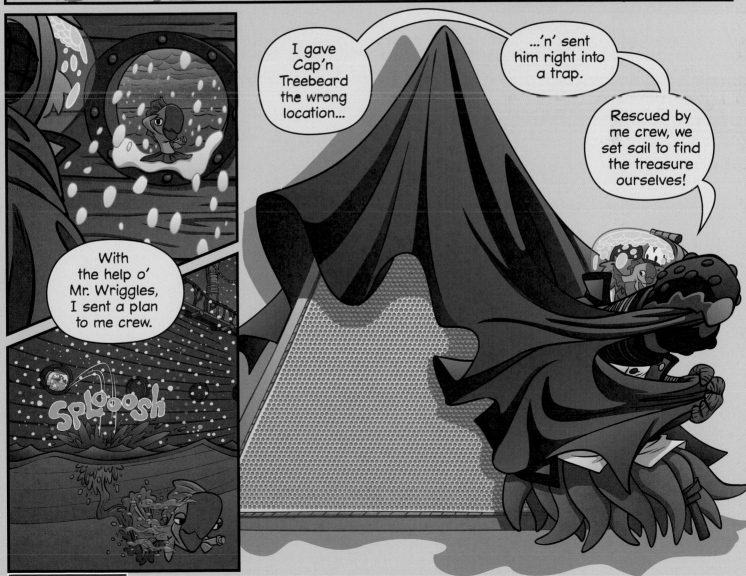

Contents: Chapter 8

See page 36 in the Practice book for a recommended reading/practice sequence for Chapter 8.

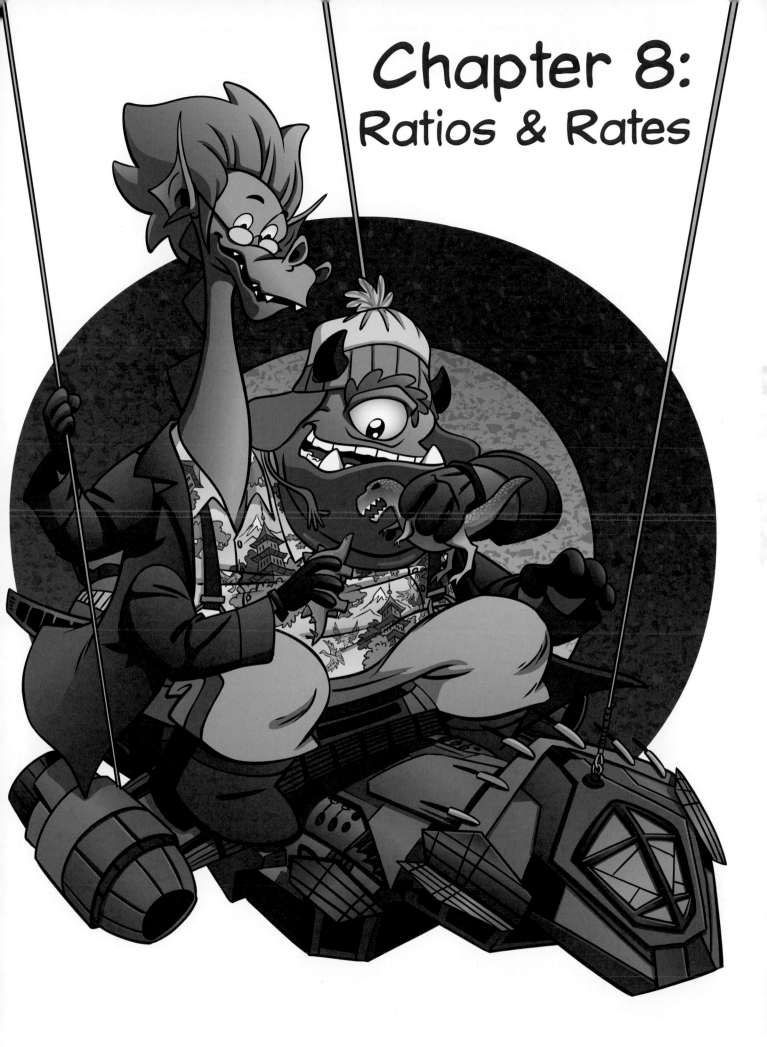

Chapter 8:
Ratios & Rates

45

There are a few different ways to write ratios.

The most common ratio notation uses a colon.

For example, the ratio of 24 ounces of soda to 60 ounces of juice is written as 24:60, which is read "24 to 60."

$$\text{soda} : \text{juice} = 24:60$$

We can also use fractions to work with ratios.

For example, if the ratio of soda to juice is 24 to 60, then the amount of soda divided by the amount of juice is $\frac{24}{60}$.

$$\frac{\text{soda}}{\text{juice}} = \frac{24}{60}$$

Like fractions, ratios are almost always written in simplest form.

We simplify ratios the same way we simplify fractions.

$\frac{24}{60}$ simplifies to $\frac{2}{5}$. So, a 24:60 ratio of soda to juice simplifies to 2:5.

$$\frac{\text{soda}}{\text{juice}} = \frac{24}{60} \xrightarrow{\div 12}{\div 12} \frac{2}{5}$$

$$\text{soda} : \text{juice}$$
$$= 24 : 60$$
$$= 2 : 5$$

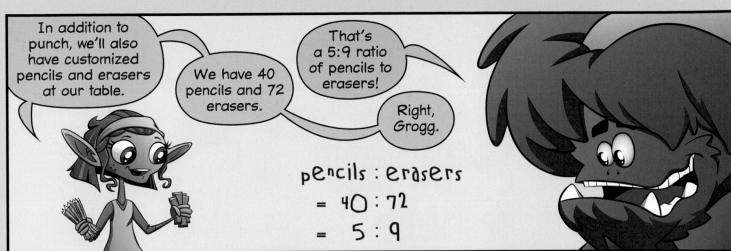

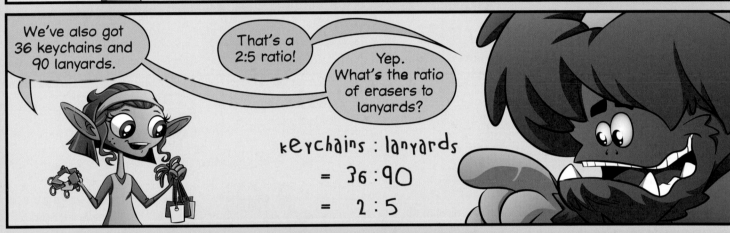

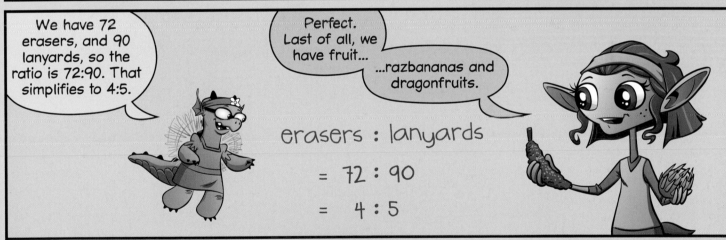

Practice: Pages 37-47

51

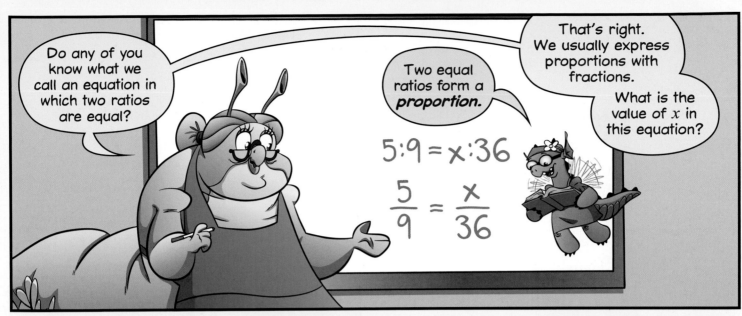

Try all three.

53

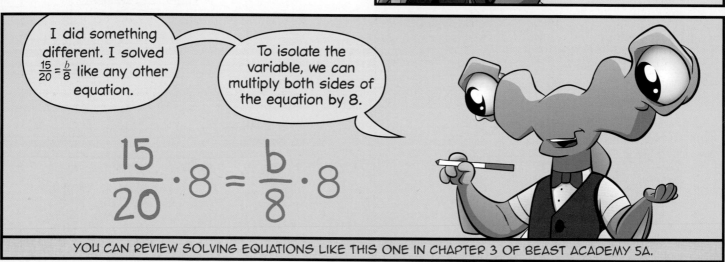
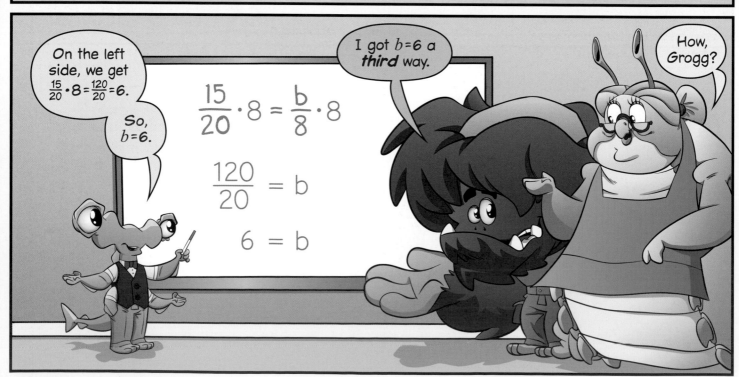

I wanted to get rid of the fractions first, so I multiplied both sides by 40.

Why 40?

$$\frac{15}{20} \cdot 40 = \frac{b}{8} \cdot 40$$

The LCM of 20 and 8 is 40, so multiplying both sides by 40 cancels the denominators on both sides.

We get $30 = 5b$.

$$\frac{15}{20} \cdot \cancel{40}^{2} = \frac{b}{8} \cdot \cancel{40}^{5}$$

$$30 = 5b$$

$$\frac{15}{20} \cdot \cancel{40}^{2} = \frac{b}{8} \cdot \cancel{40}^{5}$$

$$30 = 5b$$

$$6 = b$$

Dividing both sides by 5 gives us $b = 6$.

All fine methods, little monsters! Well done!

How can each method be used to solve *this* equation?

$$\frac{3}{5} = \frac{7}{d}$$

How would *you* solve for d?

55

We can't simplify $\frac{3}{5}$, and there isn't an integer we can multiply 3 by to get 7.

$$\frac{3}{5} = \frac{7}{d}$$

But, we can multiply 3 by $\frac{7}{3}$ to get 7.

Since $5 \cdot \frac{7}{3} = \frac{35}{3}$, we have $d = \frac{35}{3} = 11\frac{2}{3}$.

$$\overset{\cdot\frac{7}{3}}{\frac{3}{5}} = \frac{7}{11\frac{2}{3}} \quad \underset{\cdot\frac{7}{3}}{}$$

To solve $\frac{3}{5} = \frac{7}{d}$, we need to isolate d.

But, d is in the denominator.

$$\frac{3}{5} = \frac{7}{d}$$

However, if we multiply both sides by d, we remove d from the denominator!

$$\frac{3}{5} \cdot d = \frac{7}{d} \cdot d$$

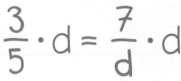

Then, we multiply both sides by 5...

...and divide both sides by 3.

$d = \frac{35}{3}$, which equals $11\frac{2}{3}$.

$$\frac{3}{5} \cdot d = \frac{7}{d} \cdot d$$
$$\frac{3d}{5} = 7$$
$$3d = 35$$
$$d = \frac{35}{3}$$
$$d = 11\frac{2}{3}$$

Practice: Pages 48-57

58

Workshop
RATES
Map Room
What are these, Captain Kraken?
These be the fastest recorded times for sailin' between various points on Beast Island.
Are any of these *your* records?
Aye, a few.
Whoa! Captain Conch sailed 51 miles in just 3 hours!?
Captain Crustacean sailed 90 miles from Cape Grindylow to Port Hydra in just 5 hours!
We call him Cap'n Crusty, for short.
Beast Island
Beast Bay
Serpent Strait

Wait...
...which of those is faster?
'Tis an excellent inquiry.
How can we be comparin' the speeds o' Captain Conch 'n' Captain Crusty?
Conch:
51 miles in 3 hours
Crusty:
90 miles in 5 hours
Who sailed faster?

What if the two captains sailed for the **same** amount of time? Since 3 and 5 are both factors of 15, we can figure out how far each captain could sail in 15 hours.

Conch:
51 miles in 3 hours
255 miles in 15 hours

Crusty:
90 miles in 5 hours
270 miles in 15 hours

Conch sailed 51 miles in 3 hours. At that speed, he could go $51 \cdot 5 = 255$ miles in $3 \cdot 5 = 15$ hours.

Crusty sailed 90 miles in 5 hours. At that speed, he could go $90 \cdot 3 = 270$ miles in $5 \cdot 3 = 15$ hours.

Crusty was faster.

Conch:
51 miles in 3 hours = 17 miles per hour

Crusty:
90 miles in 5 hours = 18 miles per hour

Instead of figuring out how far each Captain could sail in **15** hours, I figured out how far each could go in **1** hour.

For Conch to travel 51 miles in 3 hours, he had to travel $51 \div 3 = 17$ miles per hour.

For Crusty to travel 90 miles in 5 hours, he had to travel $90 \div 5 = 18$ miles per hour.

"PER" MEANS "FOR EACH" OR "FOR EVERY."

Excellent figurin', little monsters. **Speed** be the ratio of distance to time.

We often express speed as the distance traveled for one unit o' time...

...for example, as the number o' miles traveled per hour.

We can abbreviate "miles per hour" as mph.

$$Speed = \frac{distance}{time}$$

$$= \frac{90 \ mi}{5 \ hr}$$

$$= \frac{18 \ mi}{1 \ hr}$$

$$= 18 \ mph$$

Speed be a special type o' ratio, called a **rate**.

A rate describes how much o' one quantity there be for one unit of another quantity.

If ye' be usin' the word **per**, then you're probably usin' a rate.

Rate

For example, the time I faced a mighty spiralac, me heart be thumpin' 200 beats **per** minute.

Its enormous feet be havin' 5 razor-sharp talons **per** toe!

Its nostrils could spray powerful jets of water like a firehose...

...30 gallons **per** second!

What treasure were you after?

Spiralacs shed their soft, fluffy feathers.

While me crew distracted her, I was able to gather a dozen sacks full o' the downy plumage.

The largest sack o' feathers weighed just $4\frac{1}{2}$ pounds 'n' sold for $720!

The smallest sack weighed only $2\frac{1}{5}$ pounds 'n' sold for $330!

Which sack be worth more per pound?

What was the cost per pound of each sack?

To find out what a sack of feathers is worth in dollars per pound...

...we divide its price in dollars by its weight in pounds.

$$\frac{720 \text{ dollars}}{4\frac{1}{2} \text{ pounds}}$$

A $720 sack that weighs $4\frac{1}{2}$ pounds is worth $720 \div 4\frac{1}{2}$ dollars per pound.

To divide 720 by $4\frac{1}{2}$, we convert $4\frac{1}{2}$ to a fraction... $4\frac{1}{2} = \frac{9}{2}$.

$$\frac{720 \text{ dollars}}{4\frac{1}{2} \text{ pounds}} = 720 \div 4\frac{1}{2}$$
$$= 720 \div \frac{9}{2}$$

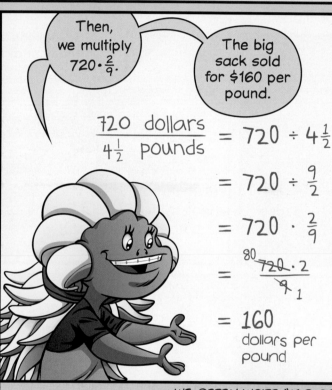

Then, we multiply $720 \cdot \frac{2}{9}$.

The big sack sold for $160 per pound.

$$\frac{720 \text{ dollars}}{4\frac{1}{2} \text{ pounds}} = 720 \div 4\frac{1}{2}$$
$$= 720 \div \frac{9}{2}$$
$$= 720 \cdot \frac{2}{9}$$
$$= \frac{\overset{80}{720} \cdot 2}{\underset{1}{9}}$$
$$= 160 \text{ dollars per pound}$$

A $330 sack that weighs $2\frac{1}{5}$ pounds is worth $330 \div 2\frac{1}{5}$ dollars per pound.

So, the small sack sold for $150 per pound.

The large sack was worth $10 more per pound!

$$330 \div 2\frac{1}{5}$$
$$= 330 \div \frac{11}{5}$$
$$= 330 \cdot \frac{5}{11}$$
$$= \frac{\overset{30}{330} \cdot 5}{\underset{1}{11}}$$
$$= 150 \text{ dollars per pound}$$

WE OFTEN WRITE "160 DOLLARS PER POUND" AS $160/LB.

Aye.

'Tis easy to compare two rates when both be in dollars per pound.

Were there any sacks of feathers that were even *more* valuable than those two?

Practice: Pages 58-63

Sure. A 96-ounce bag weighs $\frac{96}{16} = 6$ pounds.

But for more complicated unit conversions, we usually use *conversion factors*.

Huh?

A conversion factor is a fraction that equals 1.

Like $\frac{8}{8}$?

Not exactly. In a conversion factor, the units in the numerator and the denominator are different.

But... How can the fraction equal 1?

The numerator and denominator have to be equal.

For example, since 1 pound equals 16 ounces, $\frac{1 \text{ lb}}{16 \text{ oz}}$ equals 1.

We can use this conversion factor to convert any number of ounces to pounds.

$$\frac{1 \text{ lb}}{16 \text{ oz}} = 1$$

Conversion Factor

How?

How could you use a conversion factor to convert 96 ounces to pounds?

We multiply!

You can multiply anything by 1 without changing its value.

So, when we multiply 96 ounces by $\frac{1 \text{ lb}}{16 \text{ oz}}$, we aren't changing the weight...

...just the units.

$$96 \text{ oz} \cdot \frac{1 \text{ lb}}{16 \text{ oz}}$$

$$96 \text{ oz} \cdot \frac{1 \text{ lb}}{16 \text{ oz}}$$

$$= \frac{96 \text{ oz} \cdot 1 \text{ lb}}{16 \text{ oz}}$$

$$= \frac{96 \text{ lb}}{16}$$

$$= 6 \text{ lb}$$

Units cancel the same way common factors do.

I see. Ounces cancel in the numerator and the denominator, and we're left with pounds.

Yep.
$$96 \text{ oz} \cdot \frac{1 \text{ lb}}{16 \text{ oz}}$$
$$= \frac{96 \text{ lb}}{16} = 6 \text{ lb}.$$

What else do we need to buy?

Captain Kraken asked me to pick up 7 kilograms of melonberries for the Swashbucklin' Club.

Kilograms? Each of these cartons holds 20 **ounces** of melonberries.

We'll need to convert.

How?

We can use conversion factors!

There are 1,000 grams in a kilogram...

...and there are 28 grams in an ounce.*

How many ounces are in 7 kilograms?

ACTUALLY, 1 OZ = 28.3495 GRAMS, BUT 28 GRAMS IS CLOSE ENOUGH FOR MELONBERRY PURCHASES.

First, we can convert from kilograms to grams...

...then from grams to ounces.

$$kg \rightarrow g \rightarrow oz$$

$$\frac{1,000\,g}{1\,kg} = \frac{1\,kg}{1,000\,g} = 1$$

Since there are 1,000 grams in a kilogram, both of these fractions equal 1.

How do we know which one to multiply by?

Remember... We want the kilogram units to cancel.

So, we multiply by $\frac{1,000g}{1kg}$ to convert kilograms to grams.

$$7\,\cancel{kg} \cdot \frac{1,000\,g}{1\,\cancel{kg}} = 7,000\,g$$

That gives us 7,000 grams.

Now, we need to convert from grams to ounces.

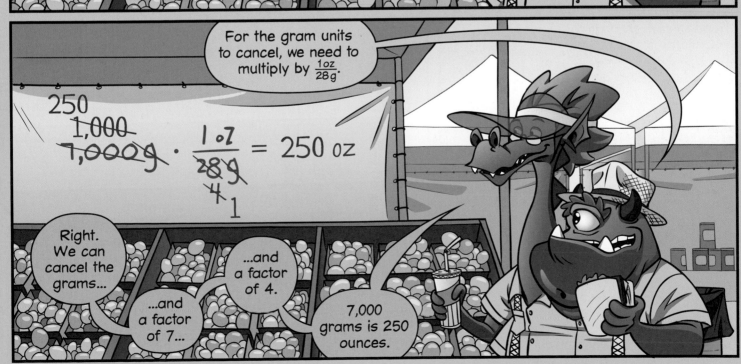

For the gram units to cancel, we need to multiply by $\frac{1oz}{28g}$.

$$\cancel{7,000}\,\cancel{g} \cdot \frac{1\,oz}{\cancel{28}\,\cancel{g}} = 250\,oz$$

(annotations: 250 over 1,000 over 7,000 g; and 4 over 28; and 1)

Right. We can cancel the grams...

...and a factor of 7...

...and a factor of 4.

7,000 grams is 250 ounces.

$$\overset{25}{\cancel{250}oz} \cdot \frac{1 \text{ carton}}{\underset{2}{\cancel{20}oz}} = \frac{25 \text{ cartons}}{2} = 12\tfrac{1}{2} \text{ cartons}$$

$$7 \text{ kg} \cdot \frac{1,000 \text{ g}}{1 \text{ kg}} \cdot \frac{1 \text{ oz}}{28 \text{ g}} \cdot \frac{1 \text{ carton}}{20 \text{ oz}}$$

Practice: Pages 64-73

Contents: Chapter 9

See page 74 in the Practice book for a recommended
reading/practice sequence for Chapter 9.

Chapter 9:
Decimals

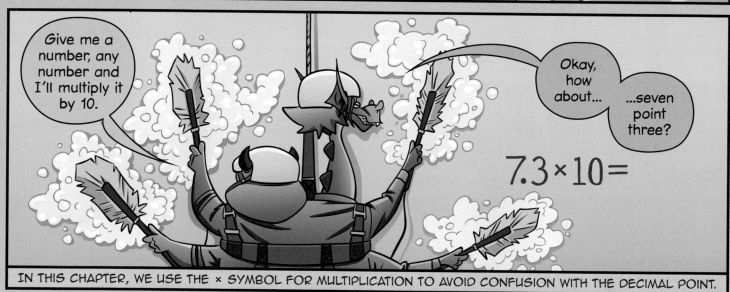

IN THIS CHAPTER, WE USE THE × SYMBOL FOR MULTIPLICATION TO AVOID CONFUSION WITH THE DECIMAL POINT.

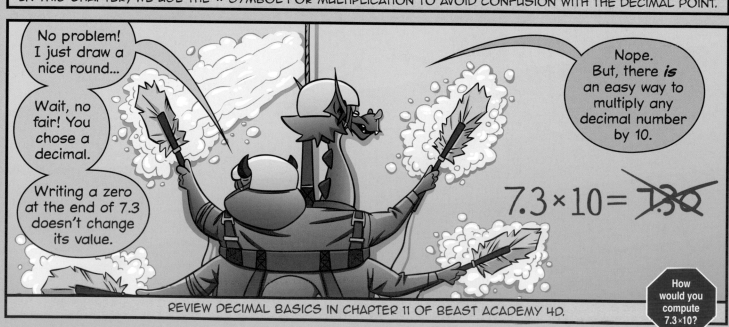

REVIEW DECIMAL BASICS IN CHAPTER 11 OF BEAST ACADEMY 4D.

$7.3 \times 10 = 7\frac{3}{10} \times 10$

$= \left(7 + \frac{3}{10}\right) \times 10$

$= (7 \times 10) + \left(\frac{3}{10} \times 10\right)$

$= 70 + 3$

$= 73$

Hmmm, don't tell me.

7.3 is $7\frac{3}{10}$.

I can multiply $7\frac{3}{10} \times 10$.

$7 \times 10 = 70$, and $\frac{3}{10} \times 10 = 3$. So, $7\frac{3}{10} \times 10$ is $70 + 3 = 73$.

That's interesting.

The 7 moved from the ones place to the tens place...

...and the 3 moved from the tenths place to the ones place.

Try multiplying another decimal by 10.

How about 0.45×10?

$0.45 \times 10 =$

I'll distribute again.

0.45 is $\frac{4}{10} + \frac{5}{100}$. When we multiply by ten, we get $4 + \frac{5}{10}$, which is 4.5.

The digits moved to the next larger place value again!

$0.45 \times 10 = \left(\frac{4}{10} + \frac{5}{100}\right) \times 10$

$= \left(\frac{4}{10} \times 10\right) + \left(\frac{5}{100} \times 10\right)$

$= 4 + \frac{5}{10}$

$= 4\frac{5}{10}$

$= 4.5$

I'll use fractions again.

5.9 is $5\frac{9}{10}$, and 0.1 is $\frac{1}{10}$.

I can use the distributive property to multiply $5\frac{9}{10} \times \frac{1}{10}$.

$5 \times \frac{1}{10} = \frac{5}{10}$, and $\frac{9}{10} \times \frac{1}{10} = \frac{9}{100}$.

$$5.9 \times 0.1 = 5\frac{9}{10} \times \frac{1}{10}$$
$$= \left(5 + \frac{9}{10}\right) \times \frac{1}{10}$$
$$= \left(5 \times \frac{1}{10}\right) + \left(\frac{9}{10} \times \frac{1}{10}\right)$$
$$= \frac{5}{10} + \frac{9}{100}$$
$$= 0.59$$

Written as a decimal, we get 0.59.

You got it. To multiply by one tenth, you can just shift each digit to the next **smaller** place value.

Since every place value is one tenth the place value to its left!

5.9×0.1
$= 0.59$

Exactly.

So, multiplying a number by 0.1 shifts its decimal point one place to the **left**.

5.9×0.1
$= 0.59$

Cool.

It gets even better!

How could you compute 3.8×0.000001?

$$3.8 \times 0.000001 =$$

Try it.

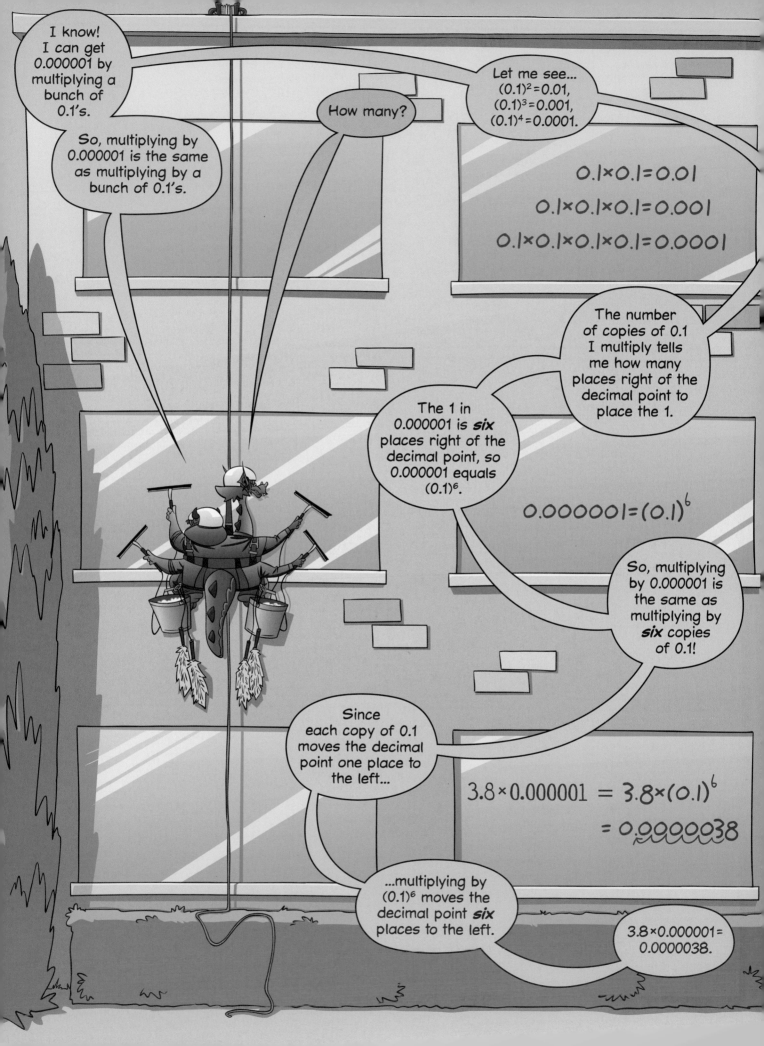

Cool, right?

The same way that multiplying by powers of 10 like 100, 1,000, and 1,000,000 moves the decimal point to the right...

...multiplying by powers of 0.1 like 0.01, 0.001, and 0.000001 moves the decimal point to the left.

$$0.541 \times 100 = 54.1$$
$$23.9 \times 1,000 = 23,900.$$
$$0.92 \times 1,000,000 = 920,000.$$

$$0.541 \times 0.01 = 0.00541$$
$$23.9 \times 0.001 = 0.0239$$
$$0.92 \times 0.000001 = 0.00000092$$

When multiplying by a power of 10, you count the zeros in the power of 10 to figure out how far to move the decimal point to the **right**...

...and when multiplying by a power of 0.1, you count the digits right of the decimal point to figure out how far to move the decimal point to the **left.**

$$0.541 \times 100 = 54.1$$
$$23.9 \times 1,000 = 23,900.$$
$$0.92 \times 1,000,000 = 920,000.$$

$$0.541 \times 0.01 = 0.00541$$
$$23.9 \times 0.001 = 0.0239$$
$$0.92 \times 0.000001 = 0.00000092$$

10^n HAS n ZEROS. FOR EXAMPLE, $10^6 = 1,000,000$.
$(0.1)^n$ HAS A 1 THAT IS n DIGITS RIGHT OF THE DECIMAL POINT. FOR EXAMPLE, $(0.1)^6 = 0.000001$.

Sometimes I get my left and right mixed up.

Not me. When I hold up my hand, it makes a convenient "L" for "left."

I don't have a left hand.

Practice: Pages 75-81

WE WILL NOT ASK YOU TO COMPUTE PRODUCTS LIKE THE ONE ABOVE.
NONE OF THE QUESTIONS WE ASK IN BEAST ACADEMY REQUIRE THE USE OF A CALCULATOR.

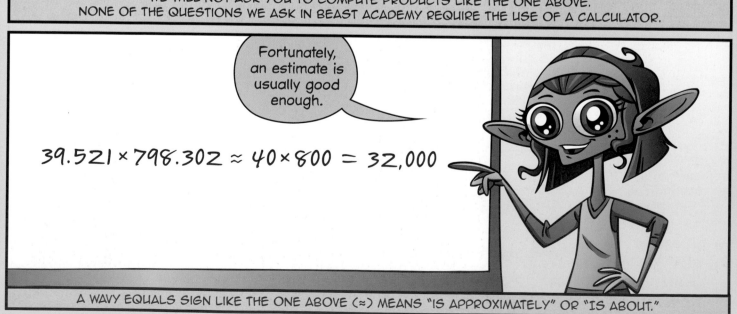

A WAVY EQUALS SIGN LIKE THE ONE ABOVE (≈) MEANS "IS APPROXIMATELY" OR "IS ABOUT."

Some decimal products are not difficult to compute.

For example, try this one.

0.007×0.03

Without the compubot.

We can write both decimals as fractions.

$0.007 = \frac{7}{1,000}$, and $0.03 = \frac{3}{100}$.

And $\frac{7}{1,000} \times \frac{3}{100} = \frac{21}{100,000}$.

$$0.007 \times 0.03$$
$$= \frac{7}{1,000} \times \frac{3}{100}$$
$$= \frac{21}{100,000}$$

As a decimal, $\frac{21}{100,000}$ is 0.00021.

0	.	0	0	0	2	1
ones		tenths	hundredths	thousandths	ten-thousandths	hundred-thousandths

$$0.007 \times 0.03$$
$$= \frac{7}{1,000} \times \frac{3}{100}$$
$$= \frac{21}{100,000}$$
$$= 0.00021$$

So, $0.007 \times 0.03 = 0.00021$.

$$0.024 \times 0.9$$
$$= (24 \times 0.001) \times (9 \times 0.1)$$
$$= 24 \times 9 \times 0.001 \times 0.1$$
$$= 216 \times 0.001 \times 0.1$$
$$= 0.0216$$

0.024 is 24×0.001, and 0.9 is 9×0.1.

24×9=216, and we move the decimal point 4 places to the left to get 0.0216.

We don't even need the middle steps!

$$0.024 \times 0.9$$
$$3 + 1 = 4 \text{ digits total}$$

We can just count the number of digits right of the decimal point in 0.024 and 0.9 to figure out where to place the decimal point in their product.

$$= 0.0216$$
4 digits after decimal

That's right!

Careful, though. What do you get when you multiply 0.125 by 0.08?

$$0.125 \times 0.08$$

Try it.

To multiply 0.125×0.08, we can start by ignoring the decimal points. 125×8=1,000.

Then, we count the total number of digits right of the decimal point in 0.125 and 0.08.

$$125 \times 8 = 1{,}000$$

$$0.125 \times 0.08$$
3   2

There are 3+2=**5** digits right of the decimal point.

So, we move the decimal point in 1,000 so that there are **5** digits right of the decimal point.

$$0.125 \times 0.08 = 0.01000$$
3   2   5

$$= 0.01$$

But, 0.01000=0.01. So, 0.125×0.08 =0.01.

Good. You have to include the trailing zeros when you place the decimal point.

But, it's fine to remove them afterwards. For example, 0.5×0.06 =0.030... ...which equals 0.03.

tee Hee Hee Hee Hee

$$0.5 \times 0.06$$
$$= 0.030$$
$$= 0.03$$

What's so funny, Grogg?

I multiplied 1.57×3.4267...

...and the compubot gave me "giggles."

GIGGLES=

1.57×3.4267...

82

Practice: Pages 82-91

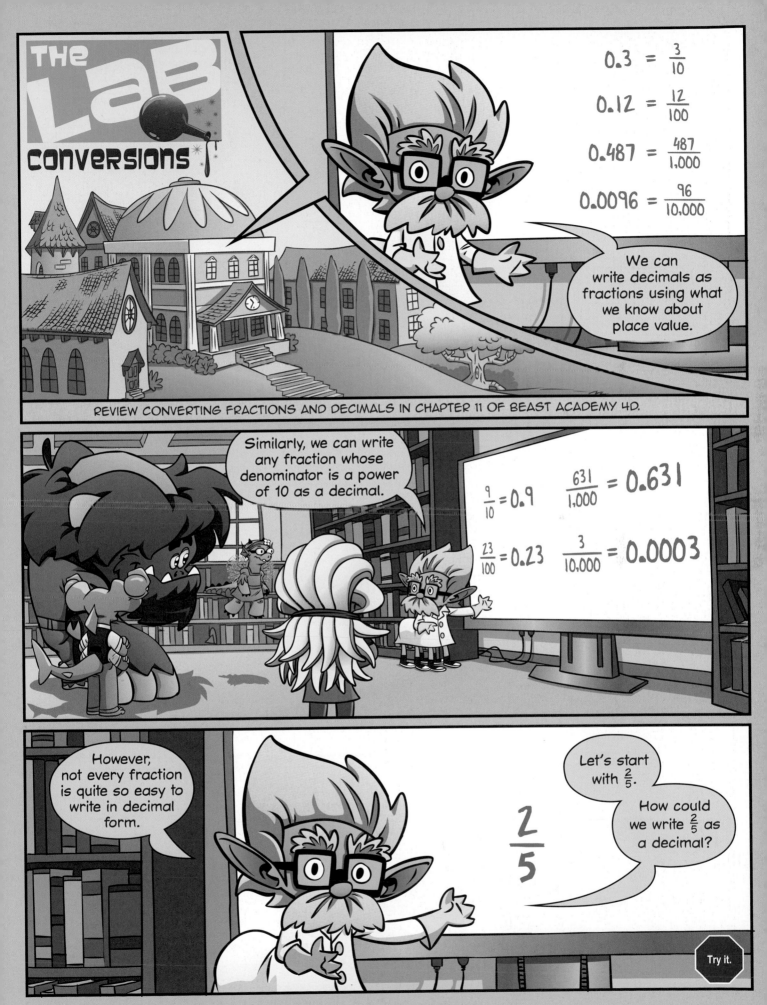

We can convert $\frac{2}{5}$ so it has a 10 in the denominator. $\frac{2}{5} = \frac{4}{10}$.

Then, $\frac{4}{10} = 0.4$.

I got the same answer a different way.

Since $\frac{2}{5} = 2 \div 5$, I divided 2 by 5.

$$\frac{2}{5} = \frac{4}{10} = 0.4$$

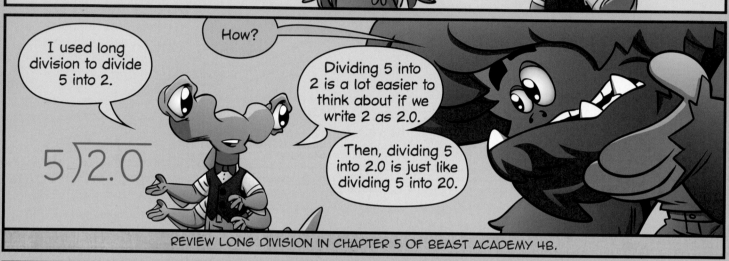

I used long division to divide 5 into 2.

How?

Dividing 5 into 2 is a lot easier to think about if we write 2 as 2.0.

Then, dividing 5 into 2.0 is just like dividing 5 into 20.

$$5 \overline{)2.0}$$

REVIEW LONG DIVISION IN CHAPTER 5 OF BEAST ACADEMY 4B.

With whole numbers, since $4 \times 5 = 20$, we know 5 goes into 20 **4** times...

...with nothing left over.

$$\begin{array}{r} 4 \\ 5\overline{)20} \\ -20 \\ \hline 0 \end{array}$$

And since $0.4 \times 5 = 2.0$, we know 5 goes into 2.0 **0.4** times...

...with nothing left over.

So, $\frac{2}{5} = 0.4$.

$$\begin{array}{r} 0.4 \\ 5\overline{)2.0} \\ -2.0 \\ \hline 0.0 \end{array}$$

Well done. Try a few more.

Convert each of these fractions into a decimal.

$$\frac{9}{20} \qquad \frac{11}{25} \qquad \frac{3}{8}$$

Try all three.

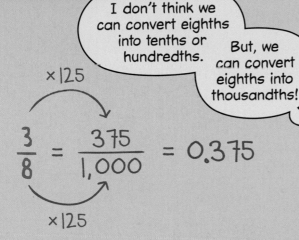

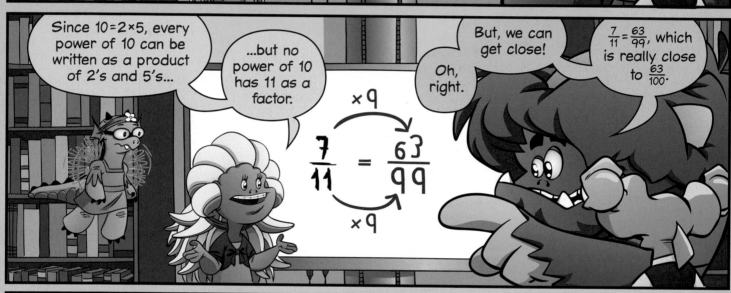

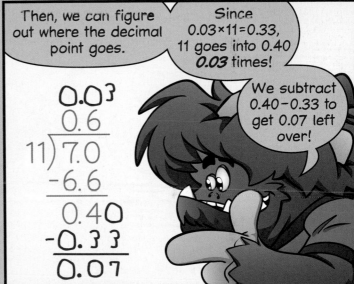

89

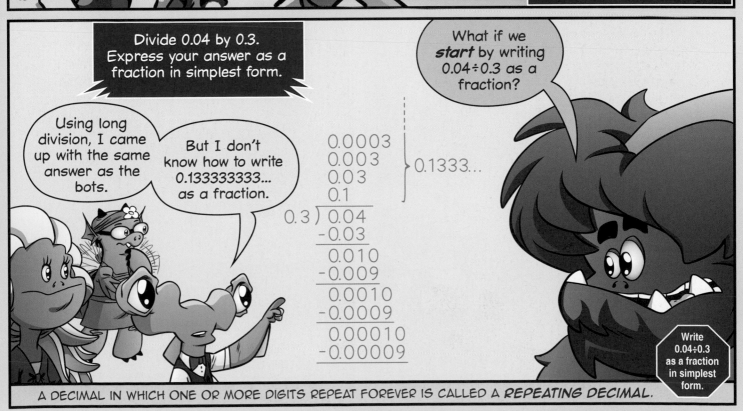

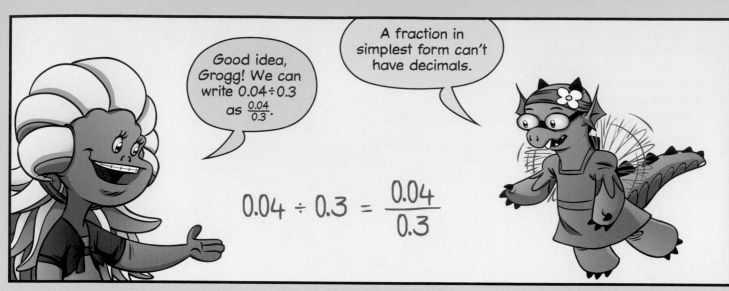

Good idea, Grogg! We can write $0.04 \div 0.3$ as $\frac{0.04}{0.3}$.

A fraction in simplest form can't have decimals.

$$0.04 \div 0.3 = \frac{0.04}{0.3}$$

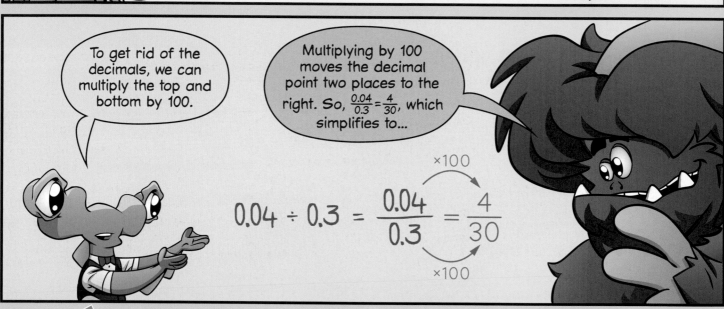

To get rid of the decimals, we can multiply the top and bottom by 100.

Multiplying by 100 moves the decimal point two places to the right. So, $\frac{0.04}{0.3} = \frac{4}{30}$, which simplifies to...

$$0.04 \div 0.3 = \frac{0.04}{0.3} \xrightarrow[\times 100]{\times 100} \frac{4}{30}$$

ding!

$\frac{2}{15}$!

$\frac{2}{15}$ is correct. The Little Monsters score the first point.

Question 2:
If A, B, and C are distinct digits such that $0.A \times 0.B = 0.C$, what is the largest possible sum of A, B, and C?

$$0.A \times 0.B = 0.C$$

Try it.

94

I found something!

First, I wrote 1.5, 0.75, and 0.6 as simplified fractions.

3, 1.5, 1, 0.75, 0.6, ___

$3, \dfrac{3}{2}, 1, \dfrac{3}{4}, \dfrac{3}{5},$ ___

I didn't see a pattern, so I wrote the whole numbers as fractions, too.

3, 1.5, 1, 0.75, 0.6, ___

$\dfrac{3}{1}, \dfrac{3}{2}, \dfrac{1}{1}, \dfrac{3}{4}, \dfrac{3}{5},$ ___

Since all of the other fractions had 3 in the numerator, I wrote $\frac{1}{1}$ as $\frac{3}{3}$.

3, 1.5, 1, 0.75, 0.6, ___

$\dfrac{3}{1}, \dfrac{3}{2}, \dfrac{3}{3}, \dfrac{3}{4}, \dfrac{3}{5},$ ___

Next in the pattern is $\frac{3}{6}$!

As a decimal, that's...

3, 1.5, 1, 0.75, 0.6, ___

$\dfrac{3}{1}, \dfrac{3}{2}, \dfrac{3}{3}, \dfrac{3}{4}, \dfrac{3}{5}, \dfrac{3}{6}$

---

ding!

0.5!

Correct!

Question 4:
What is the rightmost nonzero digit when $\left(\dfrac{3}{5}\right)^{99}$ is written as a decimal?

Try it.

ding!

6!

Correct!

$0.6^{99}=0.0000000000000000000001088864$
$3725001181768278171119300963675$
$61906184121591145257178661061582$
856912896

Whoa!

How'd you get that so fast, Winnie!?

I knew that $\frac{3}{5}=0.6$.

So, I just needed the rightmost nonzero digit of $(0.6)^{99}$.

To find $(0.6)^{99}$, we compute 6^{99} and then move the decimal 99 places to the left.

So, the units digit of 6^{99} gives us the rightmost nonzero digit of $(0.6)^{99}$.

And since multiplying any two numbers with units digit 6 gives a number with units digit 6...

...every power of 6 ends in 6, including 6^{99}!

REVIEW UNITS DIGIT COMPUTATIONS IN CHAPTER 2 OF BEAST ACADEMY 4A.

Question 5:
In a 2.75-gram coin, the ratio of gold to copper is 6:5. How many grams of gold are in the coin? Express your answer as a decimal.

How many grams?

The Little Monsters beat the Bots to the buzzer.

1.5 grams of gold!

Correct!
The Little Monsters lead 5 to 0 and have a chance for a perfect score!

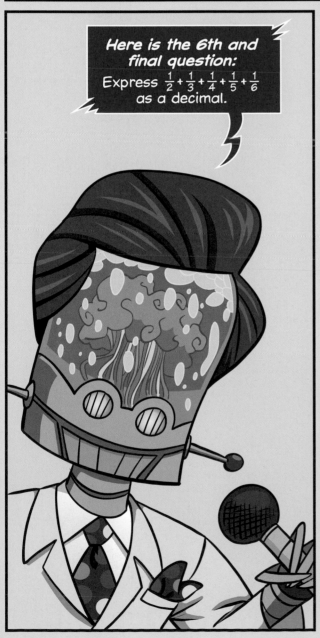

Here is the 6th and final question: Express $\frac{1}{2} + \frac{1}{3} + \frac{1}{4} + \frac{1}{5} + \frac{1}{6}$ as a decimal.

We need to be fast. The bots will have no trouble converting all those fractions to decimals.

Maybe we can add all of the fractions first, then convert the answer to a decimal.

I think I see an even faster way!

Find the sum.

99

Three of the fractions add up to 1.

$$\frac{3}{6} + \frac{2}{6} + \frac{1}{6} = \frac{6}{6} = 1$$

$\frac{1}{2}$ $\frac{1}{3}$ $\frac{1}{4}$ $\frac{1}{5}$ $\frac{1}{6}$

$$\frac{3}{6} + \frac{2}{6} + \frac{1}{6} = \frac{6}{6} = 1$$

$\frac{1}{2}$ $\frac{1}{3}$ $\frac{1}{4}$ $\frac{1}{5}$ $\frac{1}{6}$

$$0.25 + 0.2 = 0.45$$

The other two are easy to write as decimals!

$\frac{1}{4}$ is 0.25, and $\frac{1}{5}$ is 0.2.

0.2 + 0.25 = 0.45.

The sum of all five is 1 + 0.45 =

ding!

1.45!

$\frac{1}{2}$ = 0.5
$\frac{1}{3}$ = 0.333333333333...
$\frac{1}{4}$ = 0.25
$\frac{1}{5}$ = 0.2
$\frac{1}{6}$ = 0.166666666666...

0.5
0.333333333333...
0.25
0.2
+ 0.166666666666...
1.449999999999...

That answer is...

Practice: Pages 92-107

Index

Want more Beast Academy?
Try Beast Academy Online!

Learn more at BeastAcademy.com